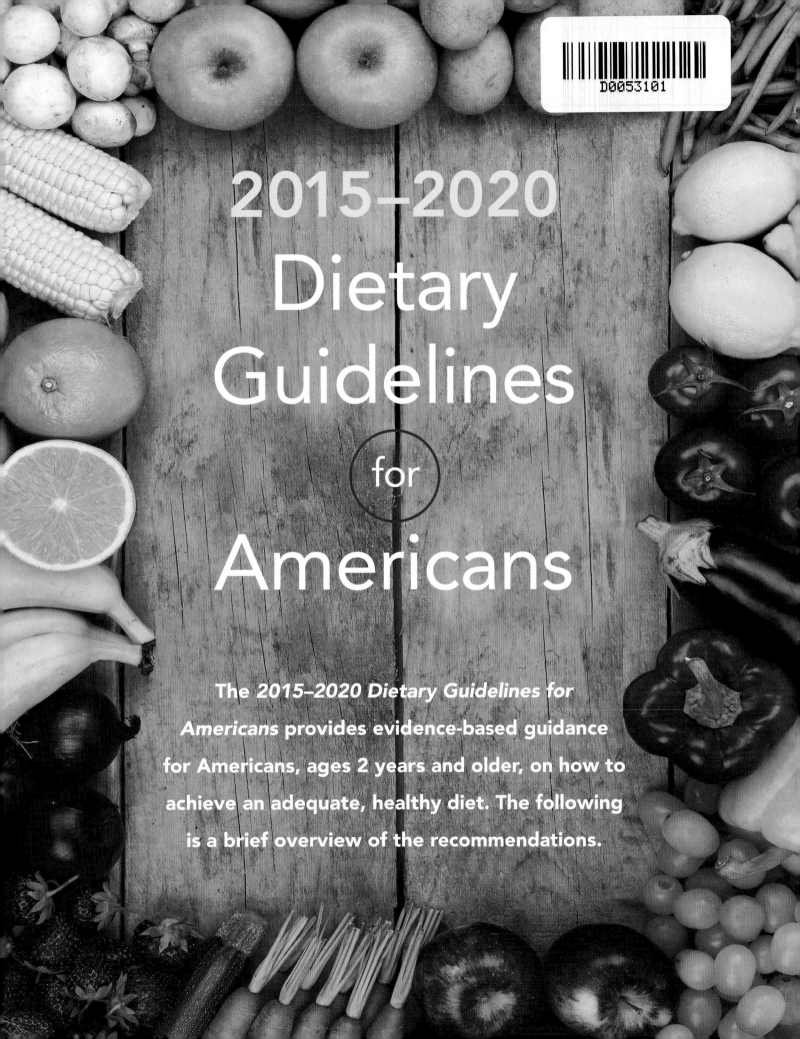

2015–2020 Dietary Guidelines for Americans

for

The *2015–2020 Dietary Guidelines for Americans* provides evidence-based guidance for Americans, ages 2 years and older, on how to achieve an adequate, healthy diet. The following is a brief overview of the recommendations.

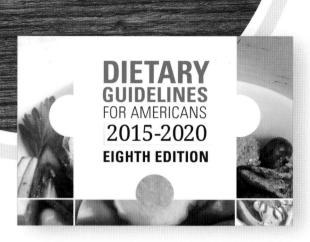

DIETARY GUIDELINES FOR AMERICANS 2015-2020 EIGHTH EDITION

The *2015–2020 Dietary Guidelines for Americans* has three primary objectives:

- **Promote health**
- **Prevent chronic disease**
- **Help people reach and maintain a healthy weight**

To achieve these objectives, five specific guidelines are highlighted with accompanying Key Recommendations.

1 Follow a healthy eating pattern

A healthy eating pattern is the sum of what an individual usually consumes that comes within set limits for total energy, saturated and *trans* fats, added sugars, and sodium. Choosing a healthy eating pattern will help you achieve and maintain a healthy body weight, support nutrient adequacy, and reduce the risk of chronic disease. Your nutritional needs should be met primarily through nutrient-dense foods, which can be fresh, frozen, or canned. In some cases, you may need to consume fortified foods and dietary supplements to provide nutrients that you otherwise might not obtain through food and beverages alone. To help you achieve a healthy eating pattern, the *Dietary Guidelines* provides Key Recommendations for which foods to include and which to limit.

TABLE 1. KEY RECOMMENDATIONS FOR FOODS TO INCLUDE AND LIMIT TO ACHIEVE A HEALTHY EATING PATTERN

A healthy eating pattern *includes*:	A healthy eating pattern *limits*:
A variety of vegetables from all subgroups • Dark green • Red and orange • Legumes (beans and peas) • Starchy • Other	Added sugars
Fruits, particularly whole fruits	Saturated and *trans* fats
Grains, at least half of which are whole grains	Sodium
Fat-free or low-fat dairy • Milk, yogurt, cheese and/or fortified soy beverages	Alcohol
A variety of protein foods • Seafood • Lean meats and poultry • Eggs • Legumes • Nuts and seeds • Soy products	
Oils	

Healthy eating patterns are adaptable and may reflect sociocultural and personal preferences. The **Healthy US Style** follows the *2015–2020*

Healthy US Style

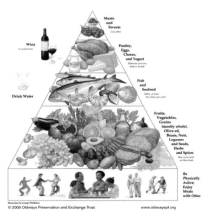

Mediterranean Style

© 2009 Oldways Preservation & Exchange
Trust, www.oldwayspt.org

Healthy Vegetarian

© 2013 Oldways Preservation & Exchange
Trust, www.oldwayspt.org

DASH Diet

Dietary Guidelines for Americans (see later in this Study Card for details on MyPlate). The **Mediterranean Style** diet is associated with reduced risk for cardiovascular disease (CVD). It contains more fruits and seafood and less dairy than the Healthy US Style. The **Healthy Vegetarian** pattern contains greater amounts of soy, legumes, nuts/seeds, and whole grains than the Healthy US Style. The **DASH (Dietary Approaches to Stop Hypertension) Diet** was designed to increase intake of foods expected to lower blood pressure (fruits, vegetables, and whole grains) while reducing sodium and empty calories, resulting in overall reduced CVD risk.

PHYSICAL ACTIVITY

Whereas a healthy diet is one component of achieving a healthy weight and reducing chronic disease risk, physical activity is also important. Accordingly, the *2015–2020 Dietary Guidelines for Americans* recommends that ALL Americans meet the recommendations for physical activity set forth in the *Physical Activity Guidelines for Americans*.

- **ADULTS:** at least 150 minutes of moderate-intensity physical activity per week, including muscle-strengthening exercises on 2 or more days.

- **YOUTH AGES 6 TO 17 YEARS:** at least 60 minutes of physical activity per day, including aerobic, muscle-strengthening, and bone-strengthening activities.

2 Focus on variety, nutrient density, and amount

What are "nutrient-dense" foods? Nutrient-dense foods are those that contain vitamins, minerals, fiber, and natural substances that have positive health effects. Examples include grilled chicken, spinach, eggs, brown rice, milk, and avocado.

What does "variety" mean? To meet nutrient needs, it is important to eat fruits and vegetables of all colors of the rainbow, vary your proteins to include those from animal and plant sources, and consume different whole grains such as popcorn, oatmeal, whole wheat, and barley.

5 Support healthy eating patterns for all

To successfully adopt the Key Recommendations in the *2015–2020 Dietary Guidelines for Americans*, everyone has a role to play. The Social-Ecological Model for Food and Physical Activity suggests that collective action is needed at home, school/work, in the community, and at food retail outlets to ensure that ALL Americans have access to healthy foods that are both affordable and familiar.

Achieving a Healthy Eating Pattern with MyPlate

MyPlate is a tool designed to make the *2015–2020 Dietary Guidelines* achievable for all Americans. It provides an image of what a healthy plate should look like to serve as a reminder that a healthy eating style is possible throughout a lifetime. Specifically, it emphasizes three points:

1. Focus on a variety of nutrient-dense foods in appropriate portion sizes to achieve nutrient needs within a specific calorie range.

2. Limit foods high in saturated fat, added sugars, and sodium.
   - Read the Nutrition Facts panel, and choose lower-sodium options for soups, breads, and frozen meals.
   - Replace sugary beverages with water

3. Start with small changes to build healthier eating styles.
   - A healthy eating pattern is a journey that is influenced by age, preferences, food access, culture, and traditions. Making small changes over time will help to achieve a healthy eating pattern.
   - Tackle one small change at a time.
   - Make half of your plate fruits and vegetables.
     - Focus on eating whole fruits instead of juice or dried fruit
     - Vary veggies.
   - Make half your grains whole grains.
   - Move to low-fat and fat-free dairy.
   - Vary your protein routine.
   - Eat and drink the right amount for you.

MyPlate can be customized to meet the energy and nutrient needs of all Americans ages 2 years and older, including pregnant and lactating women and adults with increased levels of physical activity.

To get your personalized plan and tips to achieve it, visit **www.ChooseMyPlate.gov**.

The complete guidelines and more information are available at http://health.gov/dietaryguidelines/2015/guidelines

SATURATED FATS

Saturated fats are fats that are solid at room temperature, such as butter and meat fat, and include fats found in full-fat cheese, whole milk, ice cream, poultry skin, and many baked goods. Americans consume the majority of saturated fats in mixed dishes such as pizza, burgers and meat dishes, snacks and sweets, and protein foods. The *Dietary Guidelines* recommends that all Americans limit their saturated fat intake to 10% or less of total daily calories.

SODIUM

Sodium is of particular public health concern in the United States because of its relationship with high blood pressure (hypertension) and kidney disease. Accordingly, a Key Recommendation is to consume less than 2,300 milligrams of sodium each day.

OTHER DIETARY COMPONENTS OF IMPORTANCE

Alcohol: If alcohol is consumed, it should be in moderation. For adults of legal drinking age, women should consume no more than one drink per day and men should consume no more than two drinks per day. Many adults should avoid alcohol entirely, such as pregnant women.

Trans Fats: Given its effect on blood cholesterol, Americans should keep intake of *trans* fats as low as possible by limiting intake of solid fats and foods that contain partially hydrogenated oils, which are often found in margarines and processed peanut butter.

Dietary Cholesterol: The previous version of the *Dietary Guidelines* recommended that Americans limit their intake of dietary cholesterol to less than 300 milligrams/day. The new *Dietary Guidelines* provides no specific amount but recommends consuming as little dietary cholesterol as possible.

Seafood: The *Dietary Guidelines* recommends consuming 8 ounces of seafood per week as a way to obtain beneficial amounts of omega-3 fatty acids. Pregnant women should consume at least 8 and up to 12 ounces of fish per week to promote healthy fetal development.

Caffeine: Caffeine is not a nutrient, but rather a dietary component that functions in the body as a stimulant. It is naturally found in chocolate, tea, and coffee and is added to other foods and beverages, such as soda and energy drinks. People who do not consume caffeine should not start. However, moderate consumption (400 milligrams/day) can be incorporated into a healthy eating pattern. Consumers should consider the cream, milk, and added sugars that their coffee and soda may contain.

4 Shift to healthier food and beverage choices

Replace foods and beverages that are high in calories, sugar, salt, and/or saturated fats with those that are more nutrient dense, both across and within all food groups. Examples of healthy replacements include the following:

- Chips and dip » Carrots and hummus
- Apple-flavored cereal bar » Apple
- White bread » Whole-wheat bread

- Sugar-sweetened beverages » Water
- Butter » Oils
- Whole milk » Fat-free or skim milk

These small changes can have a big impact on health over time. You can find tips for changing intake habits within each food group at **www.choosemyplate.gov/start-small-changes**.

Fruits
1 1/2 cups
1 cup of fruits counts as
• 1 cup raw or cooked fruit; or
• 1/2 cup dried fruit; or
• 1 cup 100% fruit juice.

Vegetables
2 1/2 cups
1 cup vegetables counts as
• 1 cup raw or cooked vegetables; or
• 2 cups leafy salad greens; or
• 1 cup 100% vegetable juice.

Grains
6 ounce equivalents
1 ounce of grains counts as
• 1 slice bread; or
• 1 ounce ready-to-eat cereal; or
• 1/2 cup cooked rice, pasta, or cereal.

Food group targets for a 1,800 calorie pattern are: >

Protein
5 ounce equivalents
1 ounce of protein counts as
• 1 ounce lean meat, poultry, or seafood; or
• 1 egg; or
• 1 Tbsp peanut butter; or
• 1/4 cup cooked beans or peas; or
• 1/2 ounce nuts or seeds.

What does "amount" mean? Excess weight affects two-thirds of adults and one-third of children in the United States. Thus, Americans need to pay more attention to portion sizes and the total amount of foods and drinks they consume each day.

The *Dietary Guidelines* can be personalized with food group targets for all Americans based on age, sex, and activity level using the Daily Checklist at the MyPlate website.

Dairy
3 cups
1 cup of dairy counts as
• 1 cup milk; or
• 1 cup yogurt; or
• 1 cup fortified soy beverage; or
• 1 1/2 ounces natural cheese or 2 ounces processed cheese.

3 Limit calories from added sugars and saturated fats, and reduce sodium intake

On average, Americans consume too many calories from added sugars and saturated fat and eat too much salt. Often foods high in sugar and saturated fat displace healthier foods from the diet, such as fruits, vegetables, and whole grains. Accordingly, the *Dietary Guidelines* specifically recommends that individuals limit their intake of foods with these components.

ADDED SUGARS

A Key Recommendation in the *Dietary Guidelines* is to limit added sugar intake to **10% or less of total daily calories.** For example, an individual following a diet of 1,800 calories per day can consume 180 calories from added sugars. This is equivalent to 45 grams or 11.25 teaspoons of sugar each day. Whereas that seems like a lot of sugar, Americans consume as much as 18% of their calories from added sugars. To decrease sugar intake, read labels on processed foods, especially beverages, snacks and sweets, and refined grains, which contribute the majority of added sugars to the American diet. Sugars from fruit, dairy, and unprocessed grains are considered natural and do not count toward added sugar recommendations.

ISBN-13: 978-0-13-445947-9
ISBN-10: 0-13-445947-4

9 780134 459479

90000

PEARSON   ALWAYS LEARNING